st Century Skills **INNOVATION LIBRARY**

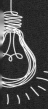

# Smart Farming

Martin Gitlin

Published in the United States of America by Cherry Lake Publishing
Ann Arbor, Michigan
www.cherrylakepublishing.com

Reading Adviser: Marla Conn, MS, Ed., Literacy specialist, Read-Ability, Inc.

Photo Credits: © Panumas Yanuthai/Shutterstock, cover, 1; © MONOPOLY919/Shutterstock, 5; © Marso Tyre/Flickr, 7; © Daniel J. Heinlin/Shutterstock, 8; © Allison McAdams/Shutterstock, 9; © Drazen Zigic/Shutterstock, 10; © Yuangeng Zhang/Shutterstock, 11; © Igor Borodin/Shutterstock, 12; ©Maike Schulz/Shutterstock, 15; © Monopoly Monopoly/Dreamstime, 17; © Kent Weakley/Shutterstock, 18; © Jevtic/Dreamstime.com, 21; © Thomas La Mela/Shutterstock, 23; © bdk/WikiMedia, 24; © Aknarin Jit Ong/Dreamstime.com, 27; Public Health Image Library/Public Domain Files, 29

Graphic Element Credits: © Ohn Mar/Shutterstock.com, back cover, multiple interior pages;
© Dmitrieva Katerina/Shutterstock.com, back cover, multiple interior pages;
© advent/Shutterstock.com, back cover, front cover, multiple interior pages;
© Visual Generation/Shutterstock.com, multiple interior pages;
© anfisa focusova/Shutterstock.com, front cover, multiple interior pages;
© Babich Alexander/Shutterstock.com, back cover, front cover, multiple interior pages

Copyright © 2021 by Cherry Lake Publishing Group

All rights reserved. No part of this book may be reproduced or utilized
in any form or by any means without written permission from the publisher.

Library of Congress Cataloging-in-Publication Data has been filed and is available at catalog.loc.gov.

Names: Gitlin, Marty, author.
Title: Smart farming / written by Martin Gitlin.
Description: Ann Arbor, Michigan: Cherry Lake Publishing, 2020. | Series: Exploring the internet of things | Includes index.
Identifiers: LCCN 2020003059 (print) | LCCN 2020003060 (ebook) | ISBN 9781534168961 (hardcover) | ISBN 9781534170643 (paperback) | ISBN 9781534172487 (pdf) | ISBN 9781534174320 (ebook)
Subjects: LCSH: Artificial intelligence—Agricultural applications—Juvenile literature. | Agricultural innovations—Juvenile literature. | Internet of things—Juvenile literature.
Classification: LCC S494.5.I5 G58 2020 (print) | LCC S494.5.I5 (ebook) | DDC 338.1/60285—dc23
LC record available at https://lccn.loc.gov/2020003059
LC ebook record available at https://lccn.loc.gov/2020003060

Printed in the United States of America
Corporate Graphics

# Table of Contents

Introduction .................................................................... 4

Chapter one
Smart Farming Devices ............................................... 6

Chapter two
Benefits and Challenges ............................................. 16

Chapter three
The Future of Smart Farming ..................................... 22

Critical Thinking and Problem Solving ...................... 28
Learn More .................................................................. 30
Glossary ....................................................................... 31
Index ............................................................................ 32

# Introduction

Farmers feed the world. Crop farmers grow fruits and vegetables. Dairy farmers raise cows that produce milk used to make yogurt and cheese. Ranchers raise cattle that satisfy the demand for meat. Poultry farmers raise chickens and turkeys. They also bring eggs to the market. Farmers have an important job. It's not an easy task. But the Internet of Things (IoT) can help.

IoT technology allows machines to be sensed and controlled remotely, all the while collecting important **data**. It connects the physical world to computer-based systems. IoT works through **sensors**. They are inserted within devices to **monitor** and control performance.

Farmers have reaped the benefits. IoT technology has helped them farm smarter and more efficiently. Farmers can connect tractors, livestock, and even soil to the internet! The result is an ability to easily monitor all aspects of the farm at once. That

High-tech farming is becoming the new industry standard.

means less wasted time, effort, and money. It also means more productive farms.

The data farmers receive from sensors give insight into which crops grow best in their soil. Sensor-driven devices can run operations such as **irrigation** systems remotely. They can even provide information that helps farmers understand what customers demand at any time.

The Internet of Things has been used on farms around the world. But it is expected to become even more popular in the future. It will continue to help farmers make their farms more efficient and valuable.

# CHAPTER ONE

# Smart Farming Devices

Did you know that agriculture has been at the forefront of the **autonomous** car industry—long before Tesla and Google? While full self-driving cars are still not 100 percent available or deemed 100 percent safe to drive on roads and highways, they are on farms and have been for quite some time. For instance, John Deere, a leading company in the agricultural industry, has been selling self-guided farming vehicles, specifically tractors, for over a decade.

IoT farming vehicles till, plow, and plant. These vehicles observe their own positions while completing a task. They decide an ideal speed for that task. They steer and brake themselves. They determine direction. They even change paths when they sense mud and **bogs**!

Self-driving vehicles are just one example of how farmers use IoT technology. Farmers and ranchers also use IoT technology

John Deere was recognized and awarded for its advanced innovation in agriculture technology at Agritechnica 2019, the largest exhibition for the agriculture industry.

There are smart wearable devices for farm animals that act like Fitbits!

to monitor the health of cows and other animals. Wearable sensors placed on an animal's collar provide details about heart rate, blood pressure, temperature, and breathing patterns.

The sensors also keep tabs on an animal's location. They can inform a farmer if one has strayed. Farmers can decide if an animal is sick by observing its grazing patterns. They relay that information to farmers, who can remove the sick animal from the rest of the herd. That prevents the spread of disease. It's incredible how large herds of livestock can be monitored simply from a tablet.

IoT technology gives crop farmers a wide range of information. Sensors provide farmers with important data that

Many smart wearable devices for farm animals can detect illness before it becomes serious.

Scientists are developing smart wearables for plants! The data collected helps farmers provide better and efficient care to their crops.

allows them to make smarter decisions. This data can then determine the best crops to grow in the current soil. They inform farmers about the ideal time and place to irrigate. Data collected from the sensors inform farmers of when their crops are ready for harvesting. Such information helps them maximize **yield**. They can even tell farmers the perfect amount of **fertilizer** and **pesticide** to use.

Farming sensors can also act as weather stations to monitor climate. This allows farmers to make more informed decisions. Among them are which crops grow best. IoT technology is also effective in a **greenhouse**. It provides more control of crop management in an already controlled environment. And it

Smart farming helps the environment as it prevents water overuse.

Precision farming allows farmers to use the resources they have more efficiently in order to produce the most crops.

allows farmers to run greenhouse irrigation and lighting systems remotely.

IoT technology plays a huge role in what is called precision farming. Precision farming allows farmers to supervise their farms remotely. It helps them monitor crop growth and health. It provides warnings of disease or **infestation**. Farmers can then take action to prevent harm to their crop output.

> The early 1990s gave birth to precision agriculture. That is when **GPS** was placed in tractors by the John Deere company. Today, this guidance system remains the most common precision farming device. It has been used extensively in the United States and around the world.
>
> Farmers benefit greatly from GPS-connected controllers in their tractors. It allows tractors to be steered automatically based on the layout of the field. That reduces steering errors by drivers. The result is less wasted seed, fertilizer, fuel, and time.
>
> IoT technology took tractor use one step further. Autonomous tractors, which require no drivers, save farmers time. And they do their jobs with great precision.

# THE BUZZ ABOUT BEEHIVES

Among those taking advantage of IoT technology are beekeepers. Beekeepers raise bees to produce honey and other products, such as royal jelly and beeswax. Or they sell the bees to scientists who study them.

Smart beehives help farmers raise healthy bees efficiently. Sensors provide data that allows them to take better care of bee colonies. One of IoT's many functions is regulating temperature in the beehive. Another is the ability to count bees as they enter and leave the colony.

Information about hive performance is sent to beekeepers through a monitor. That transmission does more than just help beekeepers keep bees healthy. It also allows them to keep an eye on their bees to prevent hive theft. Productive beehives are worth thousands of dollars.

Emerging technologies, like blockchain and IoT, are playing a major role in saving the declining bee population.

# CHAPTER TWO

# Benefits and Challenges

Farmers have used agricultural science for centuries. They gathered information that gave their farms the best chance to succeed. They examined weather reports. Crop farmers learned the best methods to plant and **cultivate** their vegetables. Ranchers studied how to care for their livestock.

But agriculture is not an exact science. Every method used for every task involves risk. IoT technology cannot eliminate risk. But it can greatly reduce it by giving far more precise data to farmers. The data collected by IoT technology could help farmers increase productivity, which could increase sales. Farmers can be certain of the best method to raise crops or care for cows. Device sensors provide this information directly to them. Sensors map and track specific data that can maximize the success of an individual farmer on any type of farm.

Also important is that IoT allows farmers to run their operations remotely. Autonomous tractors and internet-

Smart farm robots increase farm productivity in a sustainable and cost-effective way.

Smart livestock farming can help improve the health of animals.

controlled irrigation systems save farmers time and energy. And it is all done with more precision than can be achieved by humans. For instance, irrigation systems spread the precise amount of water dictated by IoT sensors. This helps avoid waste, and preserves the farmer's natural resources.

It also means cost savings. The monitoring systems for animals such as cows, chickens, and goats are one example. Sensors that help farmers remotely monitor their livestock and other animals save farmers money. That's because there is no need to hire others to supervise animals—it's already being done. The same holds true for a variety of chores around the farm. Internet-driven sensors create automation that requires no human interaction.

But the technology also comes with challenges. Smart farming is not easy. Research is vital to finding the right sensor for any tool or machine. A wrong choice results in farmers receiving the wrong type of data. Another challenge is turning good data into good decisions. That action is not always obvious based on data. Misreading what smart devices are telling them to do can prove disastrous.

Farmers must also work to protect their sensors. Sensors are often placed in fields and other outdoor areas. They can be easily damaged. Farmers trying to save money by buying cheaper sensors might find they are not durable. They must also buy smart farming **apps** made for use in the field. Farmers access data from their IoT devices on smartphones or computers. All of these devices must have enough wireless range to connect to each other.

Security is an issue as well. A smart farming app must be secure. Otherwise, data and even expensive machinery, such as an autonomous tractor, can be stolen.

IoT saves time, energy, and money. It allows farmers to monitor their farms remotely and more efficiently. But farmers must also weigh the bad with the good. Only those who learn how to best use IoT technology can truly benefit from it.

# THE HUMAN FACTOR

There is a human cost to IoT technology. People once handled the tasks that are now done by internet-connected sensors. Paid agriculture professionals once provided farmers with the information they needed. As a result, smart farms need fewer workers. The concern is that an increase in smart farming will add to the number of lost jobs.

However, IoT technology will continue to create manufacturing jobs. The industry will need more people to produce smart devices. But those same devices might result in future farms needing only one human worker: the farm owner. All other jobs might be done through automation.

History has shown that automation and technology actually create more job opportunities than the jobs they replaced.

## CHAPTER THREE

# The Future of Smart Farming

The world population keeps growing. The United Nations predicts that the human population will pass 9 billion people by the year 2050. That's more than double the population in 1980. The need for food will also grow, and the world's farms must produce that food. But nearly all the land suitable for farming is already being used. So those farms must produce more food than ever before.

This is where IoT technology can truly make an impact. Smart farming is the only way to increase yields enough to fill the need. Smart farms allow farmers to maximize production. They provide data that help farmers know when to plant, when to irrigate, and the condition of the soil.

This information alone will not ensure success in feeding the world. But farmers will no longer rely on intuition and experience to maximize their yields. They will instead rely on IoT technology to embrace precision farming methods.

According to the Food and Agriculture Organization (FAO) of the United Nations, farmers will need to double their crop production by 2050 in order to keep up with the growing population.

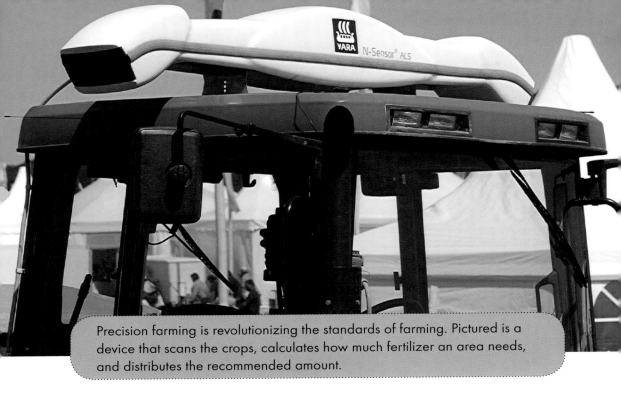

Precision farming is revolutionizing the standards of farming. Pictured is a device that scans the crops, calculates how much fertilizer an area needs, and distributes the recommended amount.

Precision farming allows farmers to fertilize where needed and plant in places they once couldn't. IoT sensors indicate the best time to plant and harvest. Automated vehicles prepare and seed the land. Early studies of smart farming show a 2 percent rise in food production.

Farmers must control diseases and pests to maintain that increase. Data from sensors allow farmers to handle outbreaks of disease and infestation—sometimes before it's even an issue! They also provide information used to prevent weeds from growing. This greatly benefits farmers. They can limit chemical spraying to certain times and problem areas.

All of these possibilities will help the environment. Pesticides and **herbicides** are major sources of pollution on farmlands.

Data that accurately predict outbreaks allow farmers to keep their crops healthy without resorting to herbicides. This contributes to the possibility of creating **eco-friendly** farms that grow safer, organic foods. Smart irrigation systems help with this too. They reduce water output by up to 70 percent, saving water and cost.

The increased use of internet-connected devices that monitor the climate will also help farmers. IoT products record all aspects of climate in small areas, such as moisture level, temperature, and humidity. What is known as **microclimate** monitoring allows farmers to improve yield and reduce wasted land.

Data that show where animals roam and where crop disease will strike will also make food safer to eat. This information allows farmers to keep their crops and animals healthy. The result will be healthier foods produced on those farms. In addition, the ability to study conditions in which disease outbreaks occur will mean a better understanding of their causes.

IoT has paved the way for entirely automated farms. The result is a brighter future for food production and the environment. That is critical given the need to feed a growing world population.

# DRONES FOR FARMS

Unmanned aerial vehicles (UAV) known as drones can be seen hovering above farmland. They are part of IoT technology. They give farmers an aerial view of their farms, which helps them fully see the lay of the land.

The drones use a series of sensors to survey specific tracts of farmland. They then transmit images that help farmers learn about soil and crop health. Drones can even spray pesticides with greater precision than tractors. This means there is less pollution being spread.

Farmers tell drones which field to survey and how high to fly before sending them on their way. The drones read conditions such as wind speed and air pressure. That allows them to determine the best flight path. They then collect visual information before landing in the exact spot from which they left.

Drones help farmers detect, eliminate, and prevent diseases from spreading among their crops.

# Critical Thinking and Problem Solving

One major drawback of IoT farming is the potential loss of jobs. It is the same problem that arose when computers took over tasks once handled by people. But training people with computer technology limited that problem. The number of people needed to program and work with computers helped the job market.

This solution might not be possible for IoT farming. Automation on all types of farms and for many chores results in fewer people needed. And the data from IoT sensors might reduce the need for paid agricultural specialists.

## Think About It

What can be done to save employment on the farms of the future?

In the past, the government invested in training and education to help workers who were displaced during the rise of automated production during the late 1960s and early 1970s.

# Learn More

## Books

Netzley, Patricia D. *Science and Sustainable Agriculture*. San Diego, CA: ReferencePoint Press, 2017.

Roberts, Josephine. *Total Tractor!* New York, NY: DK Publishing, 2015.

Rotner, Shelley. *Grow! Raise! Catch!: How We Get Our Food*. New York, NY: Holiday House, 2016.

## Websites

**Wonderopolis—What Is the Internet of Things?**
https://www.wonderopolis.org/wonder/what-is-the-internet-of-things
Learn more about IoT technology.

**YouTube—Smart Farming: IoT #1**
https://youtu.be/5mhYqb0XHxk
Watch this quick video about smart farming.

# Glossary

**apps** (APS) computer applications for mobile systems and devices

**autonomous** (aw-TAH-nuh-muhs) working without human help

**bogs** (BAHGZ) wet, spongy grounds

**cultivate** (KUHL-tuh-vate) to prepare land for growing crops

**data** (DAY-tuh) facts to be used for planning or making decisions

**eco-friendly** (EE-koh-frend-lee) not harmful to the environment

**fertilizer** (FUR-tuh-lize-ur) manure or chemical that helps soil produce better crops

**GPS** (JEE PEE ES) a system that uses satellite signals to determine a user's location and give directions; stands for global positioning system

**greenhouse** (GREEN-hous) glassed-in structure in which plants are grown

**herbicides** (HUR-bih-sidez) chemicals used to destroy plants such as weeds

**infestation** (in-fes-TAY-shuhn) presence of large numbers of insects that cause damage or disease to plants

**irrigation** (ir-uh-GAY-shuhn) the act of supplying water by artificial means

**microclimate** (MYE-kroh-klye-mit) the climate conditions in a specific area, such as a field

**monitor** (MAH-nih-tur) to check on or keep track of something

**pesticide** (PES-tih-side) substance used to destroy pests, such as insects, on crops

**sensors** (SEN-surz) detection devices that respond by transmitting a signal

**yield** (YEELD) total amount of crops harvested

# Index

agriculture. *See* farmers/farming; smart farming
Agritechnica, 7
animals, 25
    monitoring health of, 18
    and wearable devices, 8, 9
apps, farming, 19
automation, 18, 29
autonomous vehicles, 6

bees/beehives, 14, 15

chemical spraying, 24
climate monitoring, 10, 25
cost-effectiveness, 17, 18, 25
crops, 10, 12, 13
customer demand, 5

data, 4, 5, 8, 10, 14, 16, 19, 22, 25
diseases, 13, 24, 25, 27
drones, 26, 27

eco-friendly farms, 25
environment, 24

farmers/farming, 4–5
fertilizers, 10, 24
food production, 22, 23

GPS, 13
greenhouses, 10, 13

harvesting, 10, 24
herbicides, 24, 25
human costs, 20

illnesses, 8, 9
infestations, 13
irrigation systems, 5, 10, 13, 18, 25

jobs, 20, 21
John Deere, 6, 7, 13

lighting systems, 13

monitoring, 4–5, 14

pesticides, 10, 24, 26
planting time, 24
plants, 10
pollution, 24, 26
population, human, 22, 23
precision farming, 12, 13, 22, 24
productivity, 5, 16, 17, 22, 23, 24

ranchers, 4, 6, 16
research, 19
risk, 16
robots, 17

security, 19
sensors, 4, 5, 8, 10, 14, 16, 18, 19, 26
smart farming
    benefits and challenges, 16–21
    devices, 6–13
    future of, 22–29
soil, 10
sustainability, 17

temperature regulation, 14
tractors, autonomous, 6, 13, 16, 19

unmanned aerial vehicles (UAV), 26

vehicles, autonomous, 6, 13, 16, 19, 24

watering, 11
wearable devices, smart, 8, 9, 10
weather, 10
weeds, 24
workers, 20, 29

yield, 10, 22